New Ideas for a New Era

New Ideas for a New Era

Introduction to New Insights in Physics
and Mathematics for a More Complete
and Deeper Understanding of Reality.

Mustafa A. Khan, M.D.

authorHOUSE®

AuthorHouse™ LLC
1663 Liberty Drive
Bloomington, IN 47403
www.authorhouse.com
Phone: 1-800-839-8640

Published by AuthorHouse   08/19/2014

ISBN: 978-1-4969-3066-8 (sc)
ISBN: 978-1-4969-3065-1 (e)

Library of Congress Control Number: 2014914031

Table of Contents

On the consequences of a probabilistic space-time continuum.

(I) Introduction:

Our understanding of gravity has been evolving since the time of Newton. Using a spherical system of coordinates, Newton described gravitational force, at a distance r, due to a mass M, acting on a unit mass, by his famous equation: $F(M,r) = GM / r^2$, where G is the universal gravitational constant. This has worked quite well for a long time and, of course, continues to work in explaining most of the phenomenon we encounter in our everyday lives, such as calculating the trajectory of a probe to a planet within the solar system and calculating the trajectory for an artificial satellite around the earth.

The next big leap in our understanding of gravitation occurred with Einstein's General Theory of Relativity (which I will designate as GTR). In the GTR, the spatial coordinates and time were considered to be on equal footing. Instead of describing an event in a three dimensional space, (x, y, z) with time being considered a universal and absolute entity without any relation to the spatial coordinates, an event was described in a four-dimensional space-time coordinate system. With this, if we have two events, separated in space, at (x, y, z, t) and (x', y', z', t') where $x \neq x', y \neq y', z \neq z'$, then it was not necessary that $t = t'$. The GTR described the entire phenomenon equally well where Newton's theory for gravitation (hereby designated as NTG) was found to be applicable. The GTR

1

also was consistent with Bohr's correspondence principle in that it was reducible to NTG for weak gravitational fields. However, the GTR was found to be more accurate in describing phenomenon where the gravitational fields were very strong and where the NTG gave only partially correct answers, such as the precession of the planet Mercury's orbit. NTG gave an answer that was 1/2 of the actual measurement, while GTR gave an answer that agreed with the measured value almost exactly. The GTR has also been successful in describing and predicting various other phenomenon and has so far stood the test of time and experimentation. Hence, if there is to be another theory for gravitation, it will have to, as per the correspondence principle, be reducible not only to NTG, but also to GTR.

One of the limitations that have been noted very soon after the development of GTR by Einstein was that the GTR was not applicable to the atomic and sub-atomic phenomenon. The atomic and sub-atomic phenomenon is described by the Theory of Quantum Physics (henceforth referred as TQP). In TQP probability not only plays a major role but also is considered to be a characteristic of the sub-atomic world. The TQP is also consistent with the correspondence principle, as it reduces to classical physics for large masses, as it must, since classical physics has stood the test of both time and experimentation since it's formulation. The GTR does not have probability in it's description of gravitation and therefore it is unknown what phenomenon can be explained and/or predicted if one introduces a probability coordinate into the space-time continuum (hereby designated as STC) of the GTR.

In this article, I am proposing to add probability to the STC with certain characteristics and from this make certain predictions and possibly explain some of the phenomenon that have been discovered but for which a definite explanation has so far been lacking.

(II) The probabilistic space-time continuum:

We will start with the STC of the GTR, where there is no matter and where every point is fully described by the set of coordinates (x_0, x_1, x_2, x_3), (where $x_0 = t, x_1 = x, x_2 = y, x_3 = z$). We will use the short hand $\{x_i\}$, where $i = 0, 1, 2, 3$. Now, to each point $\{x_i\}$ in this STC we add a probability coordinate, P_0, and call it the baseline probability. Hence, each point in this empty STC is described by $\{x_i, P_0\}$. The probability coordinate, P_0, is as much an intrinsic characteristic of the STC as any of the x_i. This new coordinate space with probability as one of its coordinates we will call probabilistic space-time continuum (which we will designate by PSTC).

(III) The effect of matter on the PSTC:

According to the GTR, in the presence of matter each of the points $\{x_i\}$ is affected in a specific way. It is found that a mass M changes the geometry of the STC and this change in the geometry is given by a specific set of equations called the "Einstein's field equations" which connects the geometrical distortion of the STC to the matter causing the distortion. This distortion of the STC geometry by the mass M is taken to be the gravitational field of the mass M. GTR goes into details as to how objects in this distorted STC are supposed to behave and found that their behavior is similar to the behavior of a body as described by NTG due to a mass M when weak gravitational fields are considered. Just as matter affects $\{x_i\}$ it also has an effect on the probability coordinate, P_0. In the presence of matter, the P_0 "splits" into two components, P_A and P_R. P_A is the probability that an object at the point $\{x_i, P_A, P_R\}$ will have an effect that will make it move towards the mass M, while P_R is the probability that the same object at the same point, $\{x_i, P_A, P_R\}$, will have the effect that will make it move away from the mass M. Hence, in the presence of matter a point in PSTC, $\{x_i, P_0\}$ will change into $\{x_i, P_A, P_R\}$. This changing of P_0 into P_A and P_R we will call "splitting" of the baseline probability P_0. The P_0 has a baseline value of 1/2 (which I will derive later). Thus in empty PSTC each point is described by $\{ x_i, \frac{1}{2} \}$ and in the presence of matter the $\{ x_i, \frac{1}{2} \}$ "splits" into $\{ x_i, P_A, P_R \}$.

(IV) The characteristics of P_A and P_R:

To describe the characteristics of P_A and P_R, I will use a spherical coordinate system whose origin is the mass M and creating the gravitational field around it.

(1) Both P_A and P_R depend only on the distance r of the point $\{ x_i, P_A, P_R \}$ from M and not on the direction of that point, i.e. $P_A = P_R(r)$ and $P_R = P_R(r)$ for all r.

(2) Both $P_A(r)$ and $P_R(r)$ are smooth functions with respect to r.

(3) Both also functions of the mass M, i.e $P_A = P_A(M,r)$ and $P_R = P_R(M,r)$. These functions are also smooth with respect to M.

(4) For all M and r we have $P_A(M,r) + P_R(M,r) = 1$. (As a corollary

from this equation we have $\dfrac{\partial P_A(M,r)}{\partial r} = -\dfrac{\partial P_R(M,r)}{\partial r}$ for all M).

(5) $0 < P_A(M,r), P_R(M,r) < 1$, for all M and r.

(6) For a given M, there is a $r = r_0$ at which

$P_A(M,r_0) = P_R(M,r_0) = 1/2$.

(7) The functions $P_A(M,r)$ and $P_R(M,r)$ are mirror images of each other about the line $P = 1/2$ in a P v/s r coordinate frame.

(8) For $M = 0$, there is no gravitational field and therefore for an object at $\{ x_i, P_A, P_R \}$ the probability for it to move in any given direction is equal to the probability for it to move in the opposite direction, i.e. $P_A = P_R$. Since we have $P_A + P_R = 1$, this implies $P_A = P_R = 1/2$. This is precisely the baseline probability P_0, where

4

there is no matter in the PSTC. Hence, the value of $P_0 = 1/2$, as was mentioned earlier.

(9) $\text{Lim}_{r \to 0} P_A(M,r) \to 1$ and $\lim_{r \to 0} P_R(M,r) \to 0$.

(10) $\text{Lim}_{r \to \infty} P_A(M,r) \to 0$ and $\lim_{r \to \infty} P_R(M,r) \to 1$.

We can graph the functions P_A and P_R with respect to r as follows:

(1) For $M = 0$.

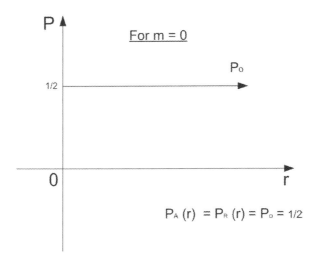

(2) For $M \neq 0$.

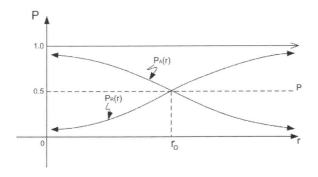

(V) Conclusions from the graph for $M \neq 0$:

Considering the graph, we can derive the following conclusions:

(1) We see that an anti-gravitational field emerges as a natural phenomenon, just as a gravitational field emerges naturally, due to the effect of matter on the probability coordinate P_0. This is due to the "splitting" of P_0 into P_A and P_R.

(2) We see that there is a distance r_0 from the mass M where $P_A(r_0) = P_R(r_0)$. This means that an object at r_0 is equally likely to move towards M, as it is to move away from M. To put it differently, at this distance r_0, an object is in a net zero gravitational field. We will call this distance, the "point of zero gravity".

(3) For $r \ll r_0$, we have $P_A(r) \approx 1$ and $P_R(r) \approx 0$.

(4) For $r \gg r_0$, we have $P_A(r) \approx 0$ and $P_R(r) \approx 1$.

(VI) Application of the probability idea to the Newton's law for gravitation:

I like to apply the above probability idea to the Newton's law for gravitation, acting on a unit mass, and expressed in spherical coordinates as: $F(M,r) = GM/r^2$, where, $F(M,r)$ is the gravitational force of attraction, on a unit mass, by mass M at distance r and G is the universal gravitational constant.

First, we have for the net gravitational force, on a unit mass, due to mass M at distance r given by:

$F(M,r) = P_A(M,r)GM/r^2 + P_R(M,r)F^*(M,r)$. Equation # 1.

Where $P_A(M,r)$ and $P_R(M,r)$ have already been defined before. $F^*(M,r)$ is the force of gravitational repulsion on a unit mass. In Newtonian language, we talk about attraction and repulsion when considering the effect of a force F on an object.

For $r = r_0$, we have,

$$F\left(M,r_0\right) = P_A\left(M,r_0\right)GM \,/\, r_0^2 + P_R\left(M,r_0\right)F^*\left(M,r_0\right).$$

But, for $r = r_0$, we have,

$P_A\left(M,r_0\right) = P_R\left(M,r_0\right)$, and the net gravitational force $F\left(M,r_0\right) = 0$.

This means: $0 = \dfrac{1}{2}GM \,/\, r_0^2 + \dfrac{1}{2}F^*\left(M,r_0\right)$.

From this we conclude, $F^*\left(M,r_0\right) = -GM \,/\, r_0^2$.

As a 1^0 approximation, and for illustrative purposes, we will take $F^*\left(M,r\right) = -GM \,/\, r^2$, for all r.

This results in the following equation for the net gravitational force on a unit mass:

$$F\left(M,r\right) = GM \,/\, r^2 \left[P_A\left(M,r\right) - P_R\left(M,r\right)\right] \text{ Equation \# 2}$$

Since $P_A\left(M,r\right) + P_R\left(M,r\right) = 1$, Equation # 2 becomes,

$$F\left(M,r\right) = GM \,/\, r^2 \left[2P_A\left(M,r\right) - 1\right] \text{ Equation \# 3.}$$

This then is the modified Newton's law for gravitation, acting on a unit mass, in spherical coordinates.

We can immediately see that for $r \ll r_0, F\left(M,r\right) \simeq GM \,/\, r^2$, (as required by the Bohr's Correspondence Principle).

Similarly, if we derive the modified gravitational field potential, in a spherical coordinate system, due to mass M at distance r, $\phi_M\left(r\right)$, we will find it to be given by the following equation:

$$\phi_M\left(r\right) = -GM\left[-\frac{1}{r} + 2\int_r^\infty \frac{P_A\left(M,r\right)}{r^2}\,dr\right]$$

(VII) Results/Discussion/Implications:

The concept of PSTC and the effect of matter on it, as has been described, lead us to entertain interesting implications.

7

(1)   As we have found that the Newton's law for gravitation is a special case for $r \ll r_0$, it is equally likely that Einstein's GTR is a special case for $r \ll r_0$. This means that we will need a modification in the GTR and obtain modified Einstein's field equations that contain P_A and P_R, which will then be applicable to all the distances within the universe. This in turn may provide explanation to certain phenomenon that has so far been difficult to explain. Also, the modified GTR (mGTR) may predict other phenomenon that has not been observed yet and which can than be searched for and see if the predications come true.

(2)   One of the phenomenon's that is known at present and needs an explanation is the existence of dark energy. From the concept of PSTC and it's interaction with matter, we can imply that dark energy is the sum total of all the anti-gravitational fields from all the visible matter $m_v(T)$ in the universe. That is:

Dark Energy $= \sum_{m_v(T)} AGF(m_v)$, where
AGF is the anti $-$ gravitational field of visible mass m_v.

   This explanation for dark energy also explains why it should have a repulsive effect on the surrounding matter.

(3)   The "problem of singularity" that plagues GTR can be resolved as follows: Given that $\lim_{r \to 0} P_A(M,r) \to 1$ and $P_A(M,r) < 1$, for all M and r, it follows that r cannot be zero.

   Since, for a singularity $r = 0$, the above constraints on $P_A(M,r)$ prevents the formation of a singularity.

(4)   The GTR has predicted the existence of gravitational waves almost a century ago. Despite the best efforts to-date, they have not been unequivocally detected. Various theories, I believe, have been put forth in the meantime to explain this. From the discussion of the effect of matter on PSTC in this article, we can put forth an explanation for the difficulty in the detection of gravitational waves. The process that produces gravitational waves is equally likely to produce anti-gravitational waves. The gravitational and the anti-gravitational waves cancel each

other out resulting in no net detectable gravitational or anti-gravitational waves.

(5) If we consider the origin of the universe from the big bang, we can at once conclude that the universe could not have started from a singularity (i.e. $r = 0$) since, as discussed above, singularities cannot exist. This means that at the time of the big bang the matter, that later gave rise to all the matter in the universe, must have had dimensions. It can be thought of as an extremely small and dense clump of some type of primordial particles. When the big bang did occur, the particles had to go through a slow phase of expansion due to $r < r_0$ of each particle relative to the others. Once the particles crossed each other's r_0 then they would go through a second phase consisting of a rapid and accelerating expansion. This two-phase expansion of the universe, at the time of the big bang, can be clearly seen from graph for $M \neq 0$ Also, from the same graph, we see that this accelerating expansion of the universe is unending due to the anti-gravitational fields produced by the visible matter and which we now see as the dark energy.

(6) We can hypothesize that the Cosmological constant, Λ, is an ad hoc representation of P_R. This means, unlike the P_R, the Cosmological constant is not a characteristic of the STC or the universe.

(7) We can express $P_A(0,r) = P_R(0,r) = 1/2$ as $\lim_{M \to 0} P_A(M,r), P_R(M,r) \to 1/2$. From this we see that the gravitational interaction between small masses is much more complicated than that given by GMm/r^2 due to significant effects from $P_R(M,r)$, even-though we are dealing with weak gravitational fields, due to the small masses, and small distances.

(8) There are likely many more implications of the ideas presented here, which I will leave it to the reader to consider.

(VIII) Conclusion:

I like to conclude this article with the following:

(1) As it is said that a good theory is one that can be disproven, one can easily test the existence/non-existence of a probability space given the modern technology. If we take two small objects of masses m_1 and m_2, with $m_1 \gg m_2$, and make m_1 the fixed object and m_2 the moving object, we should be able to find out if there is a distance r_0 at which the net gravitational force on m_2 due to m_1 is zero. If we are not able to find the r_0 for m_1, then the probability space most likely does not exist.

(2) If, however, we do find that there is a r_0 for m_1 in the above experiment, then, one can find the r_0 for different masses m and plot the r_0 against m. From this we can extrapolate to get the r_0 for bigger bodies, such as the moon, earth, sun, galaxy and so on. The practical applications of this information will clearly be many.

More thoughts on, "On the consequences of a probabilistic space-time continuum".

I) Introduction:

From our theory, "On the consequences of a probabilistic space-time continuum", we have, $\lim_{r \to \infty} P_A(M,r) \to 0$. Now we like to find the value for $\lim_{M \to \infty} P_A(M,r)$ that is consistent with the above limit and that also makes physical sense. We know that, no matter how large the mass M of an object, the $P_A(M,r)$ must approach zero as $r \to \infty$. This means that the $\lim_{M \to \infty} P_A(M,r)$ must approach zero. This is consistent both with, (1) $\lim_{r \to \infty} P_A(M,r) \to 0$ and (2) makes physical sense. From this we can conclude, given that $P_R(M,r) + P_A(M,r) = 1$, that $\lim_{M \to \infty} P_R(M,r) \to 1$. This means the more massive an object, M, the more likely it is to repel another object, m. Given our assumption that *Dark Energy* = $\sum_M AGF(m)$, where M is the total visible matter in the universe and AGF is the anti-gravitational field of a visible mass m, we can also see that the more massive an object, the bigger is its contribution to the total *Dark Energy*.

The graphs for $P_A(M,r)$ and $P_R(M,r)$ with respect to M will look as follows:

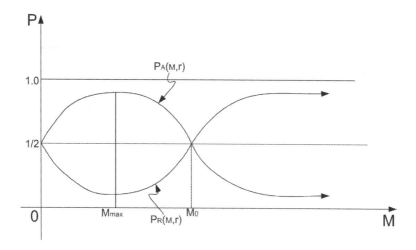

II) Conclusions:

Based upon the graph we can arrive at the following conclusions:

1) The $P_A(M,r)$ first increases with M until $M = M_{max}$, after which it starts to decline. The reverse is the case for $P_R(M,r)$.
2) There is a non-zero mass M_0 which has zero net gravitational field around it. This means we can have an object without a gravitational field!
3) Einstein's General Theory of Relativity says that a non-zero mass must produce a net positive distortion of the space-time continuum around itself. Here, we find that one can have a non-zero mass that produces no net distortion of the space-time continuum around itself.
4) From astronomical observations we should be able to find objects with $M = M_0$. If we find two stars with one orbiting the other and find that the star about which the other star is orbiting is not

being "tugged" at all, then we can say that the orbiting star has the mass of M_0.

References:

(1) On the consequences of a probabilistic space-time continuum. Mustafa A. Khan. viXra: 1401.0234. viXra.org.

On the probability of origin of the universe and other matters.

From Einstein we have the equation $E = MC^2$. We can look at this equation in a couple of ways. The first is that, any real mass M_R, also represents a "virtual energy" E_V that can be converted into real energy $E_R \left(= M_R C^2\right)$ The amount of "virtual energy" is also equal to $M_R C^2$, i.e. $E_V = M_R C^2$. Similarly, any real energy E_R represents a "virtual mass" M_V which is equal to E_R / C^2.

If we take a hyperspace (*HS*) where the laws of physics that apply to our universe come together and in which our universe exists, we can conclude that there is a probability $P(E_V)$ with which they will give rise to "virtual energy" E_V. This then can result in real mass $M_R \left(= E_V / C^2\right)$. This M_R is the original amount of matter that eventually gives rise to all the matter/energy in the universe. This matter M_R can be either ordinary matter or anti-matter, but not both. However, given the physical laws that are operating within this universe, it is entirely possible for some anti-matter to arise under certain conditions only. Also, this mass M_R is not a singularity, but has spatial dimensions. It may be extremely dense, but is not dimensionless because existence of a singularity is prohibited by the theory, "On the consequences of a probabilistic space-time continuum". We can assume, for the sake of simplicity, that it is spherical with radius *r*, but that need not be the case. It may have some irregularities with a certain maximum dimension r_{max} and a certain minimum dimension r_{min}.

From the theory, "On the consequences of a probabilistic space-time continuum", we have that there is a non-zero probability P_R that the different primordial particles (PP) of this mass M_R can repel each other. At this stage the probabilistic space-time continuum is totally wrapped around M_R within the *HS*. If we assume that this M_R is of a critical mass M_{RC} such that when the different 'PP' of it repels each other they have the right amount of momentum, μ_C, so that the 'PP' pass the point r_0 (a concept from the theory, "On the consequences of a probabilistic space-time continuum") for each other, then they will continue to repel each other with increasing probability. This then results in a rapid expansion of the universe or the probabilistic space-time continuum. From this, we see that initially there will be a slow expansion of the probabilistic space-time continuum, followed by a rapid expansion once the 'PP' cross each other's r_0. In other words, the Big Bang will consist of two stages. The first one being slow followed by a fast and accelerating second phase. For this to happen we need gravity to "de-couple" (and be greater than the attractive effects of the remaining Unified force) from the Unified force, which may occur with a certain probability $P(G)$. Also, there has to be a certain probability $P(M_{RC})$ for the critical amount of mass M_{RC} to arise so that the above events can take place. This means that the $P(E_V)$ has to be $P(E_{VC})$. Obviously the $P(E_{VC}) = P(M_{RC})$.

This means the probability of origin of our type of universe $P(O) = P(M_{RC})P(G)P_R$. Using quantum physics and the phenomenon of particle-antiparticle pair, each of mass m (i.e. total mass of $2m$), arising from empty space (thereby creating a negative energy point of $-2mc^2$) we should be able to estimate the $P(M_{RC})$. From the modified General Theory of Relativity (which is obtained using the theory, "On the consequences of a probabilistic space-time continuum") we should also be able to estimate P_R and from the Unified Field Theory (obtained using the modified General Theory of Relativity) we can get an estimate for $P(G)$. Hence, we should be able to get a numerical value for $P(O)$.

For $M_R \ll M_{RC}$, the $'PP'$ from the matter in the universe that results will most likely repel from each other in a mono-phasic "fizzle" that it would not result in a universe like ours, but will consist of a uniform "soup" of $'PP'$ or elementary particles. For $M_R \gg M_{RC}$, the Big Bang never occurs because the likely hood for the first phase to occur is so small due to $P_A \gg P_R$ (where P_A is the probability of attraction of the $'PP'$ towards each other) that the μ (momentum) of the $'PP'$ *is* not high enough to pass the r_0 of each other. Therefore, there may be only a small range for M_{RC} (similar to the goldilocks zone for our solar system) for a universe like ours to come into existence.

From the discussion so far, it is clear that the laws of physics are primary and do not require a material (matter/energy) substrate to exist, while matter/energy need the physical laws for their existence. Also, once a universe comes into existence, what occurs "within" it is completely determined and constrained by the physical laws that gave rise to it. Now, it is obvious that physical laws are just pure thoughts. Also, it is quite possible that the physical laws that operate in our universe represent only a portion of a whole universe of physical laws. This means that for our type of universe to come into existence there has to be a certain probability *P (PL)* that our physical laws have to come together. Also, the *HS* in which our type of universe exists may be one of a whole universe of hyperspaces. This then means that for our type of universe to come into existence there is a probability *P (HS)* that the right *HS* combines with the right set of physical laws, i.e. *P (U) = P (HS) P (PL)* where *P (U)* is the probability for our type of universe to come into existence. Then the actual probability for our type of universe to come into existence, *P (A)* is equal to *P (O) P (U)*. We have found that we can determine (or estimate) *P (O)*, but it is most likely not possible to find *P (U)* and hence *P (A)*.

Now, the *HS* is also a pure thought. This is so because it is a totally empty space. It is similar to the number zero which is just a mental concept without actual material existence. There is nothing with which one can represent zero, by itself, in reality. It is defined only in relation to the presence of something. Similarly, empty space is defined only in relation to the presence of matter/energy. If there is no matter/energy anywhere then there is no empty space either. The physical laws interact with *HS* to give rise to "virtual energy" E_V.

This "virtual energy" results in real matter of mass $M_R \left(= \dfrac{E_V}{C^2} \right)$ and real probabilistic space-time continuum for our type of universe. Therefore, the origin of our type of universe is the result of interaction of pure thoughts. This means what existed before the Big Bang that gave rise to our universe was pure thoughts only.

References:

(1) On the consequences of a probabilistic space-time continuum. Mustafa A. Khan. viXra: 1401.0234. viXra.org.

On a general theory of gravity based on Quantum Interactions.

I) Introduction:

Sir Isaac Newton was the first scientist to propose a formal theory for gravitation. It is given by the famous formula, written in spherical coordinate system, for a mass M at a distance r, acting on a unit mass, from the center of mass M, as $F\ (M,r)\ =\ GM\ /\ r^2$. However, he was was always plagued by the idea of "action at a distance" as he had no explanation as to how the mass M causes an effect at distance r instantaneously.

After him came others who developed the idea of a gravitational field that surrounds the body of a mass M. This helped in removing the "action at a distance" conundrum, but introduced several others. The first one being how does the mass produces this field in the first place? The second one being why does the mass produces this field? The third one being, if we were to suddenly make $M = 0$, then how does the gravitational field, that theoretically extends to an infinite distance, vanish in an instant? What process transfers that information from the object to infinity instantaneously, making the field also disappear at the same instant when $M = 0$. There are other such philosophical reasons that bothered the scientists with regards to the real existence of a gravitational field.

Einstein proposed the latest attempt for a theory of gravitation in 1915. He threw away the old idea of a gravitational field and introduced a new idea, which consisted of a warping of the space-time continuum (STC) about a mass M and due to the mass M. He went on to develop a new set of field equations that were based on the curvature (called the metric) of the STC and used Tensors instead of Vectors, which were used to describe the older field idea. This theory, The General Theory of Relativity (GTR), proved quite successful in describing the large scale structure of the universe and predicted several events, which came out to be true. One such was the curving of a beam of light while passing a large object. The other was on the precession of planet Mercury's orbit, which was exactly as predicted. The third one was the prediction about existence of black holes that has also been confirmed. Also, the GTR was similar to the Newtonian theory of gravity for weak gravitational fields as required by Bohr's Correspondence Principle (CP). But, unfortunately, even the GTR is not free from problems. One of the biggest problems is that it is totally silent about how the mass M interacts with the STC to cause the warping of the STC about it. The warping of the STC occurs right from the edge of the mass and theoretically extends to infinity. The problem as to how this distortion takes place from the point where the mass exists to infinity is very similar to a similar problem we encountered with the old gravitational field idea. Again, as with the old field idea, there are other philosophical issues that also plague GTR, such as matter "acting" on nothingness and causing a distortion of this nothingness.

There is one unique problem that faces GTR, unlike the older field idea, and this deals with its relation with Quantum Physics (QP). QP has been extremely successful in describing events and also in it's predictions about events at the microscopic scale. It is likely to be the physics that will one day supplant all other theories, including GTR, and be the paradigm for the "Theory of Everything". The problem is this: GTR is very successful on macroscopic/cosmic scale, while QP is very successful at microscopic/atomic/sub-atomic scale. The GTR idea of a STC, determinism and distortion of the STC has no correlate in QP. This is very problematic. These deep philosophical rifts between GTR and QP has been unbridgeable so far, that I am aware of, and also seem to be unbridgeable in the near future. The scientists are of the opinion that the GTR will either need to be reformulated or abandoned altogether for a different theory of gravity that is compatible with QP. This is so, because, the QP seems to be a more correct theory that describes reality than the GTR. Therefore a theory of gravity that makes QP it's bedrock and still is able to describe the macroscopic phenomenon (as well as the GTR) will be a welcome theory. In this article I like to propose such a theory of gravity.

II) A new definition for object that is compatible with QP:

For the purposes of discussion we will take an object that is spherical with mass M. Also, unless otherwise noted, we will use the spherical coordinate system centered at M.

In QP, the concept of an object is very vague. It can be taken as a particle or a wave of infinite length depending upon the type of experiment or observation one is making. This concept was proposed by De Broglie and has been proven experimentally. However, for macroscopic objects the mass is considered to be compact and has a definite position and speed. This is due to the extremely short wavelength for macroscopic objects as given by the De Broglie equation $\lambda(m,v) = h / mv$.

Here I like to introduce a concept for a mass M of any size (microscopic/macroscopic), which is different than that of De Broglie's but not incompatible with QP. Any mass M is never localized to a particular point/region of space but is smeared throughout the universe. However, this smearing of the mass follows a strict rule for the amount of mass per unit volume, $\rho_M(r)$, i.e. density, depending upon the distance from the center of the mass M. Superficially it appears to be similar to De Broglie's concept of an object, but fundamentally it is quite different from it. The exact equation for $\rho_M(r)$ is given by $\rho_M(r) = \lambda K M \left(1 - e^{\wedge}\left(-\alpha K / r^{\wedge} 2\right)\right)$, where,

1) M = Total mass of the object.
2) $K = \left(1 - e^{\wedge}\left(-\beta M\right)\right)$ and $0 \le K < 1$.
3) α, β, λ are universal constants with appropriate units.

This equation for $\rho_M(r)$ can be graphed as follows:

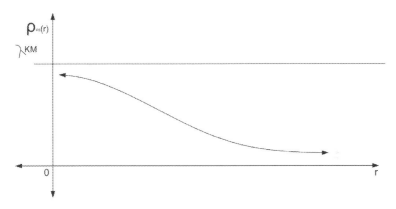

and $K = \left(1 - e^{\wedge}\left(-\beta M\right)\right)$ can be graphed as below,

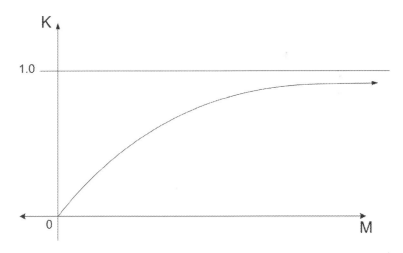

The question of why the $\rho_M(r)$ has to have this particular distribution will be explained later. Suffice it to say for now that it is due to the quantum interactions between $\rho_M(r_1)$ and $\rho_M(r_2)$.

(III) Characteristics of $\rho_M(r)$:

Considering the equation for $\rho_M(r)$ and the figure # 1, we can conclude the following properties for $\rho_M(r)$.

1) $\text{Lim}_{r\to 0}\,\rho_M(r)\to \lambda M\left(1-e\wedge(-\beta M)\right)=\lambda KM.$

2) $\text{Lim}_{r\to\infty}\,\rho_M(r)\to 0.$

3) For uniform spherical objects, $\rho_M(r)\neq\rho_M(r,\theta)$ i.e. the density of matter is the same in all directions for any given r.

4) For M, where $K\simeq1$, we have $\rho_M(r)\simeq\lambda M\left(1-e\wedge(-\alpha/r\wedge2)\right)$. From this we have $\lim_{r\to 0}\rho_M(r)=\lim_{r\to 0}\lambda M\ (1-e\wedge(-\alpha/r\wedge2)=\lambda M$. This means that for large objects where $K\simeq1$. it can be assumed that the density of matter infinitesimally close to the center is proportional to the total mass of the object. This is similar to the Newtonian concept of the "center of mass". Newton took this to be intuitively true, but here we can prove that his intuition was not wrong. However, it should be borne in mind that this approximation applies only for specifically defined large objects and certainly not for the objects dealt with in QP. Also, just as it is true for the Newtonian center of mass that the entire mass is not actually concentrated at it, so also, here we find that only λM of the total M is actually concentrated infinitesimally close to $r=0$.

5) The "gravitational force" at distance r is proportional to $\rho_M(r)$ and for well defined large objects, it is proportional to M and inversely proportional to r^2. This is simply the Newton's law of gravity, and what CP demands for new theory to be considered an improvement over an older theory. Let us consider a large M, such that we have $K\simeq1$, then $\rho_M(r)\simeq\lambda M\left(1-e\wedge(-\alpha/r\wedge2)\right)$. Taking $\left(1-e\wedge(-\alpha/r\wedge2)\right)$ and assuming $r\geq R$, where R is the "radius" of the large M and $\left|R^2\right|\gg|\alpha|$, we have as a 1^0 approximation, $\left(1-e\wedge(-\alpha/r\wedge2)\right)\simeq\alpha/r^2$. This means for large M with "radius" R and $\left|R^2\right|\gg|\alpha|$, we have $\rho_M(r)\simeq\alpha\lambda M/r\wedge2$. Since we assumed $F(M,r)\propto\rho_M(r)$, this implies $F(M,r)\propto\alpha\lambda M/r^2$ or

$F(M,r) \propto M / r^2$. This is Newton's law for gravitation and is in agreement with CP. But it should be kept in mind, once again, that we have assumed M to have specific characteristics and therefore we cannot say that the Newton's law for gravitation is applicable for all masses, especially those that are dealt with in QP. From above, we can also conclude that $G = \in \alpha\lambda$, where c is the proportionality constant whose value is given by $|c| = \left| G \middle/ {\alpha\lambda} \right|$ and where G is the universal gravitational constant.

IV) On the quantum interaction that underpins the equation:

This entire theory with the mass being smeared throughout the universe has a specific quantum interaction between $\rho_M(r_1)$ and $\rho_M(r_2)$ that gives rise to the density of matter distribution equation $\rho_M(r) = \lambda KM\left(1 - e^{\wedge}(-\alpha K / r^{\wedge}2)\right)$.

This quantum interaction I like to symbolize by I_t. It has the following form and graph.

$$I_t(\delta) = I_{max}\left(1 - \delta \middle/ {\delta_{max}}\right), \text{ where, } \delta = \text{distance between } \rho_M(r_1) \text{ and}$$

$\rho_M(r_2)$.

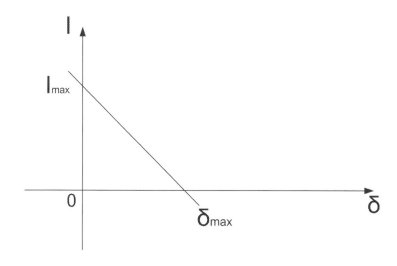

Like distances in QP, the δ_{max} is also a very short distance when compared to distances at macroscopic level and essentially zero when compared to distances at cosmic level. I also like to point out at this time that the equation $I_t\left(\delta\right) = I_{max}\left(1 - \dfrac{\delta}{\delta_{max}}\right)$, is a 1^0 approximation of a more general QI equation. But for our purposes here we will use the 1^0 approximation equation only.

When matter, i.e. $\rho_M\left(r\right)$, is coming together for the first time, then around $r = 0$ there will be a lot of $\rho_M\left(r\right)$ at distances $\delta < \delta_{max}$. As the matter accretes around $r = 0$ and the spherical mass takes shape, there will be less and less $\rho_M\left(r\right)$ after a distance R, which can be taken to be the "radius" of the object. After this point the distance δ between $\rho_M\left(r_1\right)$ and $\rho_M\left(r_2\right)$ approaches δ_{max} and when $\delta = \delta_{max}$, the $I_t\left(\delta_{max}\right) = 0$. The final shape of the object will consist of a central collection of matter followed by a gradual, smooth drop in the matter density, $\rho_M\left(r\right)$. Though theoretically the smearing can go on for infinity, in reality, it cannot because there is only a finite amount of matter to form an object. This is the reason that the "gravitational force", though can theoretically extend to infinity, in reality becomes negligible/non-existent after a certain r_{max}. To obtain the density distribution equation, $\rho_M\left(r\right) = \lambda KM\left(1 - e^\wedge\left(-\alpha K / r^\wedge 2\right)\right)$, we will need to know the exact function represented by I_{max}. One interesting question that arises here is what exactly is present at $r = 0$ that starts the process of the accretion of matter as described above, since the density of matter distribution equation never actually can have $r = 0$? The entity present at $r = 0$ cannot be visible matter nor is it dark matter. I propose to call it $\aleph$ (Aleph). This means at the heart of all objects made up of visible matter there is an entity called $\aleph$ What this $\aleph$ is we will not discuss at this time except to say that, (1) it is neither visible matter nor is it dark matter, (2) it has an attracting effect on visible matter. Perhaps I will speculate on it in another paper.

V) Conclusions:

At this point we can come to certain conclusions based upon the theory here proposed:

1) We see that, at least regarding gravity, QP provides the final explanation at all levels, from the sub-atomic to the cosmic.
2) There is no "action at a distance" which plagued Newton and his theory of gravity.
3) There is no need for any warping of the STC due to matter as required by Einstein's General Theory of Relativity.
4) As a corollary to # 3, the mysterious reason by which matter causes distortion of STC is eliminated.
5) There is no such entity as a "gravitational field".
6) Gravitational "force" is reduced to quantum mechanical interactions, I_t, at all levels, from the sub-atomic to the cosmic.
7) This is no such thing as a "gravitational interaction", in reality, between two masses, m_1 and m_2. What we have is QI between $\rho_{m_1}(r_1)$ and $\rho_{m_2}(r_2)$.
8) The gravitational acceleration $\vec{g}$ is a representation of I_t and a convenient concept to make calculations easy at macroscopic level.
9) The bending of light by a mass M is not due to the bending of the surrounding STC by M, but due to the, (1) the quantum scattering effect on the light beam and (2) the gradient of $\rho_M(r)$ that is present around an object. This will be clearer with the following diagram:

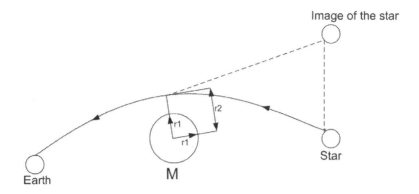

The bending of the light beam from the star is due to the combination of (1) the quantum scattering effect on the light beam in the region $(r_2 - r_1)$ and (2) the gradient of $\rho_M(r)$ in the region $(r_2 - r_1)$ with $\rho_M(r_1) > \rho_M(r_2)$.

10) For a singularity to exist, the limit of $\rho_M(r)$ as r approaches zero must be equal to infinity. However, from the matter density distribution equation we have the limit of $\rho_M(r)$ as r approaches zero being equal to $\lambda M(1 - e^{-\beta M})$, which is not equal to infinity unless M is equal to infinity. Of course, there is no object in the universe with M equal to infinity, and hence singularities cannot exist.

11) Theoretically, there cannot be any absolute empty space(s) within the universe. This is prevented by our equation, $\rho_M(r) = \lambda KM(1 - e^{\wedge}(-\alpha K / r^{\wedge} 2))$.

VI) Experiment to prove/disprove the theory:

In order for a theory to be taken seriously, one should be able to disprove it. Here I describe a simple experiment that can be done using modern technology.

Let us take an invisible laser beam and set it up as follow:

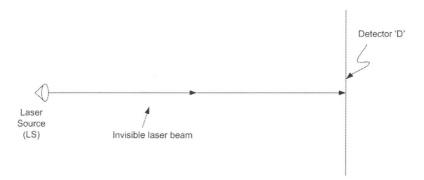

Now we bring two small masses, each of mass *m*, close to the laser beam but perpendicular to it and from the opposite directions as follows:

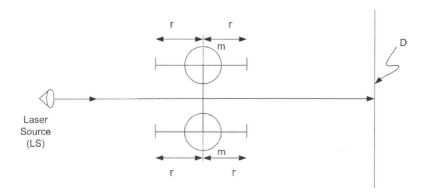

If our theory is correct, then we should see scattering *S* of the laser beam within the region 2*r*. Also, the intensity of the scattering will be greatest in the middle and falls off in either direction due to the gradient of the $\rho_m(r)$ of each mass. If our theory is correct, then, the intensity of scattering v/s distance from the center should look as follows:

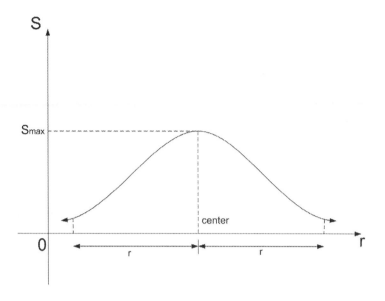

If on the other hand we get the following graph,

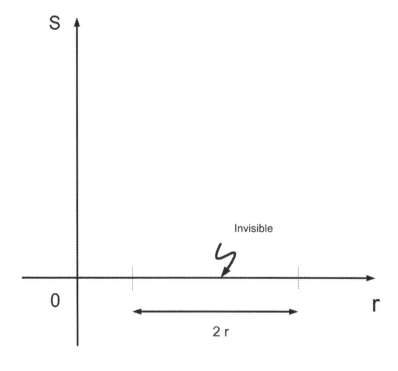

then the theory proposed here is likely to be incorrect. Of course, the laser beam will remain invisible, as there will be no scattering effect on it.

Addendum:

1) $I_t(\delta) = I_A(\delta) + I_R(\delta)$, where $I_A(\delta)$ and $I_R(\delta)$ are the attracting and repelling QI at distance δ, respectively.

2) If we assume $I_t(\delta)$ is given by $I_t(\delta) = I_{max}e^{-\delta/\delta_{max}}$, then as 1^0 approximation for $\delta < \delta_{max}$ we get $I_t \simeq I_{max}\left(1 - \dfrac{\delta}{\delta_{max}}\right)$.

3) The total QI, $I_T(r)$, on $\rho_M(r)$ is given by

$$I_T(r) = \oiiint_V^V I_t(\delta)\,dV = 4\pi \int_o^{\delta_{max}} \delta^2 I_t(\delta)\,d\delta, \text{ where } V = a$$

solid sphere of radius δ_{max} centered at $\rho_M(r)$.

Further thoughts on, "On a general theory of gravity based on Quantum Interactions". Part One.

1) In this theory, the mass M is strictly defined by the famous mass/energy equation by Einstein, $M = E / C^2$. This is different from the definition of mass as defined by Newton's first law of motion.

2) A consequence of #1 is that the inertial mass given by $M_I = F / a$, where F = force acting upon M and a is the acceleration of M_I is equal to the gravitational mass given by $M_g = F_G / g$, where F_G is the gravitational force acting on mass M_g and g is the gravitational acceleration. In short, $M_I = M_g$. This is also, of course, Galileo's "Principle of Mass Equivalence".

3) Defining mass as E / C^2, automatically converts the "matter density" equation, $\rho_M(r) = \lambda K M (1 - e^\wedge(-\alpha K / r^\wedge 2))$, with $K = (1 - e^\wedge(-M\beta))$ into an "energy density" equation given by, $\rho_E(r) = \lambda / C^\wedge 2 K_E E(1 - e^\wedge(-(\alpha K_E)/r^\wedge 2))$, with $K_E = (1 - e^\wedge(-\beta E / C^\wedge 2))$. This makes it quite easy to understand how the "matter density" equation is applicable to all objects, from the sub-atomic to the cosmic.

References:

(1) On a general theory of gravity based on Quantum Interactions. Mustafa A. Khan. viXra: 1405.0315. viXra.org.

Additional thoughts on, "On a general theory of gravity based on Quantum Interactions". Part Two.

In order to obtain practical results from the theory we can consider the "matter density" equation defining an object in terms of a fluid. We can then use the mathematics of fluid dynamics and fluid mechanics. To derive some results we will have to use the "matter density" equation and for others the equivalent "energy density" equation. Considering an object as a fluid mass leads us to consider the quantum interactions (QI) (which is a function of $\rho_M(r)$, as being equivalent to the viscosity of the fluid. Putting it in another terms, we can say that the "gravitation field strength" is equivalent to the viscosity of the fluid.

Here I am going to present a qualitative discussion on the various "gravitational" effects we can deduce using this fluid analogy for an object:

1) The bending of light due to an object which was ascribed to gravity: Using the fluid analogy, we can easily see that the bending of a beam of light by an object is simply the result of Snell's law for refraction of light. This can be diagrammed as follows:

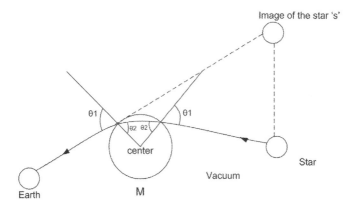

2) It is an established fact that the speed of light within a fluid is less than it's speed in vacuum, i.e. $C_{fluid} < C_{vacuum}$. Hence, if we have an object (which we have decided to consider as a fluid mass) with $\rho_M(r)$ such that the speed of light when traveling through it becomes zero, then such an object will be same as a "black hole". Of course, unlike the Einstein's General Theory of Relativity (GTR), there is no "rip" in the space-time continuum (STC) here.

3) Using fluid mechanics we should be able to calculate the orbit of a mass m about the center of mass M, i.e.

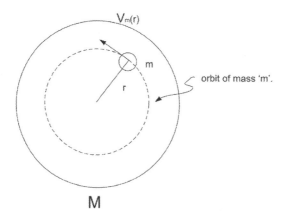

Here, M is a fluid mass with matter density of $\rho_M(r)$ at distance r through which the fluid mass m is traveling with a speed $v_m(r)$ in an orbit about the center of M. Now, considering the graph for $\rho_M(r)$

versus r (which we have seen in the paper on, "On a general theory of gravity based on Quantum Interactions), we can divide it into three sections. For the first section we have $r \leq r_1$ where the average $\rho_M(r)$, i.e. $\langle \rho_M(r) \rangle$ is $\simeq \lambda KM$. This is independent of r, which means the orbital speed, $v_m(r)$, for objects at distance r is the same for all $r \leq r_1$. The second section consists of $r_1 < r \leq r_2$, where we have the $\langle \rho_M(r) \rangle$ decrease linearly atleast to a 1^0 approximation. This means, the orbital speed $v_m(r)$ of objects at distance r, where, $r_1 < r \leq r_2$, decreases, to a 1^0 approximation atleast, linearly with r upto $r \leq r_2$. The third section consists of $r > r_2$ where the $\langle \rho_M(r) \rangle \simeq 0$, which leads to the conclusion that the orbital speed, $v_m(r)$, for objects at distance $r > r_2$ is zero, i.e. there cannot be orbits with a radius $> r_2$.

4) The time (or more precisely "Einstein time") dilation effect due to gravity:

Here we will use a clock that Einstein used to derive his Lorentz-Einstein transformation equations in his Special Theory of Relativity (STR). Our clock consists of a contraption that consists of two perfectly reflective mirrors between which a photon is bouncing back and forth between them. This clock will look something as follows:

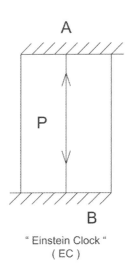

A

P

B

" Einstein Clock "
(EC)

'A' and 'B' are perfectly reflective mirrors. 'P' is the path of a photon reflecting between 'A' and 'B' without losing any energy with each reflection.

If we put this EC at different r within an object, we can immediately see that the speed of the photon is dependent on r due to $\rho_M(r)$.

This means the speed of the photon at r, C_r, is less than C_{vacuum}. This then leads to the conclusion that $\Delta t(r) > \Delta t$ (vacuum), or the time interval for the photon to travel between 'A' and 'B' is longer the closer it is to the center of the object compared to the interval when the clock is in vacuum. Also, if $r_2 > r_1$, then $\Delta t(r_1) > \Delta t(r_2)$. This means our theory of gravity, which uses three dimensional space, also allows for the time dilation effect similar to Einstein's GTR without the need for distortion of a four dimensional STC.

At this point I like to go into a brief discussion on the concept of "time". We can agree that "time" is an abstract entity. We humans have an innate intuitive feeling about the existence of this abstract entity called "time". Over the course of human history we have attempted to represent and measure this abstract entity. We can start with Galileo's attempt. He reduced time to the motion of a pendulum, which in turn is connected to gravity. This Galilean Time (GT) remained in scientific use until Isaac Newton introduced the "Newtonian Time" (NT) in his equations. This NT was also universal time. At any moment it was the same at any point in space within the universe. It was independent of motion, gravity or any other entity. It came closest to the representation of the abstract entity of time. Einstein then introduced his own representation of time, which we will call the "Einstein Time" (ET). It is based on the movement or speed of light as seen in his STR and the Einstein Clock (EC). However, to define "simultaneity" he used the movement of the hands of a mechanical clock (MC). The question whether ET as measured by EC is same as that measured by MC was never clarified in his STR. We will not go into a discussion about that either here as it will take us beyond what I like to present. Einstein showed that ET was not universal unlike NT, but that it depended on the motion of bodies and on gravity in his theories STR and GTR respectively. Recent experiments using atomic clocks (which use vibrating atoms, and hence motion) do indeed behave like ET as predicted by STR and GTR. The question that now arises is, "Does the proof of the existence of ET necessarily and sufficiently disprove the existence of NT"? I am not so sure the answer is yes. In Quantum Physics (QP) we encounter entangled particles and to accurately describe the events involving such particles (which, by the way, may be at the ends of the universe) we may have to resort to some kind of Universal Time or atleast bring back NT into the equations that describe such events. The other question is, does ET represent the abstract entity "time" accurately and completely? What about time that is based on entropy or the shortening of the telomeres on the chromosomes or the decay of living things or the psychological time we all experience and gives rise to phenomenon such as deja' vu and jamais' vu? Can we say that they are also encompassed by ET and should behave as per STR and GTR? I believe that "time" is a much broader than what Einstein thought and that ET is a gross oversimplification. The reality may be such that for some events we will need ET and for some NT and for others a completely new definition of "time". I do not think that the abstract entity "time" can be accurately and completely be reduced to any particular representation.

The next step is to make predictions based on the theory of gravitation proposed here that can not only be tested but also cannot be explained by any other theory on gravity available today.

References:

(1) On a general theory of gravity based on Quantum Interactions. Mustafa A. Khan. viXra: 1405.0315. viXra.org.

A method to determine the
shape of our Milky Way.

It may come as a surprise to people, but we do not exactly know the shape of our Milky Way! The best current estimate is that it is a barred spiral galaxy with two to four spiral arms (depending on what one considers to be an arm). We have not sent any probe that took a picture of our galaxy from above its plane and showed what our home galaxy looks like. It is also unlikely that we will ever be able to take a picture of our galaxy, any time soon, by sending a space probe that travels perpendicular to the plane of our galaxy to a distance from where it can take a picture of our Milky Way in all its glory. Hence, we need to resort to indirect methods that tell us accurately what is the exact shape of our galaxy. For this, I like to put forward a method based on the theory of gravitation that was proposed, namely, "On a general theory of gravity based on Quantum Interactions".

We have seen that our gravitational theory is based on the matter density equation, which allows us to use the mathematics of fluid mechanics and fluid dynamics. We have seen that we can apply our matter density equation to objects from the sub-atomic to the macroscopic, but there is no inherent restriction in applying it to an entire galaxy.

Let us start with a mass M that is spherical to which we apply our matter density equation. Now, let us rotate this mass. We will start to notice that the mass starts to bulge in the middle turning from a sphere into an ellipsoid. As we continue to increase the rotational speed of this mass it will start to change its shape further and starts to gradually take a shape where there is a central bulge around which is a disc of matter. Eventually, when the rotational speed reaches a certain value, there will be a vortex at the center around which will be a bulge from which will emanate spiral arms. The analogy of a fully formed hurricane will not be incorrect at this time. The galaxy at this moment will consist of a galactic eye around which will be a dense matter at an extremely high angular speed following which will be spiral arms of various lengths. Most of the matter in this galaxy will be moving in a single main plane with the matter of the central bulge moving in planes that are parallel to the main plane. Thus in our galaxy here, all the stars will be moving either in the main plane or in planes that are parallel to the main plane.

We can easily apply the above idea to find the shape of our Milky Way. Through astronomical observations we can determine the angular speed of the stars in the central bulge and from that we can find if their orbital speed is high enough to first make our Milky Way a spiral galaxy and secondly if it is a barred spiral galaxy. We should also be able to determine how many arms it has and what their lengths are. If we find stars in the center of our galaxy that are moving in orbits that are at an angle to the main plane of our galaxy, then we will have to assume that at the very center of our galaxy is not just the galactic eye but also some object that is making those stars move in orbits that are at an angle to the plane of our galaxy. Using the orbital speed of those stars we will than need to find the mass of the object that is present within the galactic eye and correct the orbital speed of the stars, moving within the galactic plane or parallel to the galactic plane, accordingly.

Thus with the use of our matter density equation and the mathematics of fluid mechanics and fluid dynamics we can not only determine the shape of our own galaxy but also explain the shapes of other galaxies. We can predict the number of shapes a galaxy can take and why and compare our predictions to the astronomical observations. This will provide yet another way to test our gravitational theory, with its matter density equation, as well as our assumption for using the mathematics of fluid mechanics and dynamics.

References:

(1) On a general theory of gravity based on Quantum Interactions. Mustafa A. Khan. viXra: 1405.0315. viXra.org.
(2) Further thoughts on, "On a general theory of gravity based on Quantum Interactions". Part Two. Mustafa A. Khan. viXra: 1406.0028. viXra.org.

A relativistic theory based on the invariance of Newton's second law for motion and the constancy of the speed of light in vacuum.

I) Introduction:

Let me begin by saying that I did not start to develop a new relativistic theory in order to somehow supersede Einstein's Special Theory of Relativity. My intention was to find out, if possible, a relationship between the Newtonian Time and Einstein Time after I discussed the concept of Time in my paper, "Further thoughts on, "On a general theory of gravity based on Quantum Interactions", Part Two". After obtaining the said relationship, I noted that I could develop a set of relativistic transformation equations between two non-Newtonian inertial reference frames, S and S' with S' moving at a constant speed, Γ, along the $+x\text{-}axis$ of S. These equations turned out to be quite different from the Lorentz-Einstein equations in Special Theory of Relativity and had consequences that in some cases were qualitatively similar to those of Special Theory of Relativity and others that were very different both qualitatively and quantitatively.

II) Newton's second law for motion:

I do not have to spell out what the Newton's second law of motion is, as it is quite well known. But I will state it's mathematical formulation, which is given by, $\vec{F}_0 = \dfrac{d\vec{P}_0}{dt_0}$. Here, (a) $\vec{F}_0$ is the Newtonian force, (b) $\vec{P}_0$ is the Newtonian momentum and (c) t_0 is the Newtonian Time. Underlying this equation are the assumptions, (1) there exists an absolute Newtonian inertial reference frame, which we will designate by S_0 and whose coordinates are (x_0, y_0, z_0) and which is at absolute rest, and (2) there exists an absolute Newtonian Time which is designated by t_0 associated with S_0. Now, both S_0 and t_0 are fictional entities and there is nothing in reality that represents them. If we take a non-Newtonian inertial reference frame S in which we use the Einstein Time which we will represent by the letter t, and whose coordinates we represent by (x, y, z) then the Newton's second law for motion, in this frame of reference, will be given by $\vec{F} = \dfrac{d\vec{P}}{dt}$.

III) Relationship between t_0 and t:

Let us have S move along $+x_0$-axis of S_0 at a uniform absolute speed 'u' relative to S_0. Using our assumption of the invariance of the second law of motion we have $\dfrac{dP_0}{dt_0} = \dfrac{dP}{dt}$ (equation 1) Now, we will assume that $dt_o = \lambda dt$ and $dP_0 = \lambda dP$. Integrating the time differential equation and putting the integration constant equal to zero so that the clocks are synchronized when $t_0 = t = 0$, we get, $t_0 = \lambda t$. The relationship between x_0 and x will be given by $x = x_0 - ut_0$. From this we get $\dfrac{dx}{dt} = \dfrac{dx_0}{dt} - u\dfrac{dt_0}{dt}$. Taking $t_0 = \lambda t$ we get $dt_0 = \lambda dt$ and putting it in the above equation we get $\dfrac{dx}{dt} = \lambda\dfrac{dx_o}{dt_0} - \lambda u$. Now, putting v_0 for $\dfrac{dx_0}{dt_0}$ and v for $\dfrac{dx}{dt}$, we get after simple manipulation $\lambda = \dfrac{v}{v_0 - u}$ (equation 2).

The equation 2 is valid for all v and v_0, including the speed of light. However, for the speed of light we have according to our assumption of it's constancy, $v_0 = v = c$. This leads to $\lambda = \dfrac{1}{\left(1 - \dfrac{u}{c}\right)}$. From this we see that the relationship between Newtonian Time and Einstein Time is given by $Newtonian\,Time = Einstein\,Time / \left(1 - \dfrac{u}{c}\right)$. We know that u is a fictional quantity, which we can never determine, which makes determination of Newtonian Time not possible.

IV) The set of relativistic transformation equations between two non-Newtonian inertial reference frames S and S':

Let us now assume that we have two non-Newtonian inertial reference frames, S and S' that are moving at uniform absolute speeds v and v' respectively relative to S_0 along the $+x_0 - axis$. We can immediately see that, $x + vt_0 = x_0 = x' + v't_0$ and $y_0 = y' = y$, $z_0 = z' = z$. After some simple manipulation we get the following set of transformation equations from S to S': $x' = x + \left(\dfrac{v - v'}{c - v}\right)ct$, $y = y'$, $z = z'$ and $t' = \left(\dfrac{c - v'}{c - v}\right)t.$

Let us now put Γ for $\left(\dfrac{v' - v}{1 - \dfrac{v}{c}}\right)$ into the above equations. This will give us the following set of Relativistic Transformation Equations from S to S': $x' = x - \Gamma t$, $y' = y$, $z' = z$ and $t' = \left(1 - \dfrac{\Gamma}{c}\right)t.$

From the consideration of these equations we can conclude, (1) the absolute and fictional quantities are eliminated, (2) the set of equations relating the coordinates (x', y', z') to (x, y, z) is exactly the same if we consider S' moving along $+x\text{-}axis$ of S at a uniform, relative to S, speed of Γ, (3) we can from this point on totally ignore the S_0 and t_0 as they have become irrelevant, (4) the speed, Γ, is a measurable speed and not something fictional, (5) there is no upper limit to the value for Γ, and (6) one can easily find out that if the speed of light in S is c, then it is also c in S'.

V) Conclusions:

1) Using symmetry we get the following transformation equations from S' to S: $x = x' + \Gamma t$, $y = y'$, $z = z'$ and $t = \left(1 + \dfrac{\Gamma}{c}\right)t'$.

2) Suppose we have a rod of length, l', relative to S', lying along the $+x'\text{-}axis$ of S'. Then given that the speed of light is the same in both S *and* S' and $\Delta t' = \left(1 - \dfrac{\Gamma}{C}\right)\Delta t$, we get, $l' = \left(1 - \dfrac{\Gamma}{C}\right)l$, which is similar to length contraction we have encountered in Einstein's Special Theory of Relativity.

3) Taking the equation $t' = \left(1 - \dfrac{\Gamma}{c}\right)t$, we get $\Delta t' = \left(1 - \dfrac{\Gamma}{c}\right)(\Delta t)$.

This is similar, qualitatively, to the time dilation we encounter in the Special Theory of Relativity. As mentioned before, there is no, a priori, upper limit on the value for Γ. Hence, if $\Gamma = c$, then $\Delta t' = 0$. This means the Einstein Time in S' (which can be a spaceship with passengers) comes to a halt. If, however, $\Gamma > c$, the $\Delta t' < 0$. This means when S' is moving at supra-luminal speed, the Einstein Time in S' moves backwards! This means if S' is a spaceship with people, then the people will get younger! It should be kept in mind that this retrogression of time is restricted to the spaceship only and the people in it will never be able to go back into the history of the universe in general or the earth in particular. The question as to whether we can build a spaceship that can travel at luminal and supra-luminal speeds is a purely engineering one. Our theory does not prohibit a spaceship from reaching any speed. One possible way to construct such a spaceship and power it was discussed by me in a paper called, "An interesting, but not practically impossible, application of the two proposed theories on gravity by myself". In it I show how we can use anti-gravity to drive a vehicle and be able to achieve any possible speed.

4) The famous mass/energy equation by Einstein, $E = MC^2$ is valid for our relativistic theory also. This is so, because, Einstein himself showed using a thought experiment involving a box and a photon in empty space that the mass/energy equation is independent of his relativistic transformation equations. Hence, it stands to reason that the mass/energy equation is also independent of our relativistic transformation equations.

5) From the equation, $dP = \lambda' dP'$, where $\lambda' = \left(1 + \dfrac{\Gamma}{C}\right)$, and the equation, $v = \dfrac{v' + \Gamma}{1 + \dfrac{\Gamma}{C}}$, where v' is the speed of an object relative to S' and v is the speed of the same object relative to S, we obtain $m(v)\left(v - \dfrac{\Gamma}{C}\right) = \left(1 + \dfrac{\Gamma}{C}\right) m'(v') v'$. We obtain the following

equations from the above equations:

a) Relativistic mass, $m(v) = m_0 \left(1 + \dfrac{v}{c}\right)^2$, where m_0 is the rest mass of the object.

b) Relativistic Kinetic Energy, KE_R, is given by $m_0 v^2 \left(\dfrac{1}{2} + \dfrac{4v}{3c} + \dfrac{3v^2}{4c^2}\right)$.

c) The total energy, E_T, that is contained in an object with relativistic mass, $m(v)$ traveling at a relativistic speed, v, is given by: $E_T = m(v)c^2 + KE_R$.

d) If we consider the above equation relating v and v', we see that for $v' = 0$, $v = \Gamma / \left(1 + \dfrac{\Gamma}{C}\right)$ and $\neq \Gamma$, as we had expected based on our knowledge from observing non-relativistic events. However, if, $\Gamma \ll C$, then $v = \Gamma$ as we expect for non-relativistic speeds. We also see that, as $\Gamma \to \infty$, $v \to c$ and not infinity as we would have expected. This is another example where we see our expectations, based on non-relativistic events, break down when we consider relativistic events.

IV) Testing the theory:

I will give two experiments that can easily be done to validate the theory presented in this paper:

(1)　If we take an electron and accelerate it to such a speed that its mass becomes equal to that of a muon, then we can expect such an "electron" to decay just like a muon. Of course, we will have to take into account the time dilation effect on the decay time of the muon. If we do not see our "electron" decay like a muon, then we can say that, either there is a flaw in our theory or more likely, that there is a more fundamental difference between our "electron" and a muon that we have not yet discovered.

(2)　If we take a photon of energy E and slow it down to zero speed, then we should see the photon give up $3/4^{th}$ of its energy in the form of radiation and its wavelength increasing four times.

References:

(1)　Further thoughts on, "On a general theory of gravity based on Quantum Interactions", Part Two. Mustafa A. Khan. viXra: 1406.0028, viXra.org.

(2)　An interesting, but not practically impossible, application derived from the theories on gravity presented by myself. Mustafa A. Khan. viXra: 1406.0032, viXra.org.

The general law of conversion
of matter and energy.

(I)

From the paper on, "Further conclusions based on the theory, "A relativistic theory based on the invariance of Newton's second law for motion and the constancy of the speed of light in vacuum", we found that when we accelerate an object with rest mass, m_0, from rest to speed v, then, (1) its mass increases from m_0 to $m_0 (1 + v / c)^2$ and (2) it will have a kinetic energy equal to $m_0 v^2 (1/2 + 4v/3c + 3v^2/4c^2)$. From (1) we see that energy can convert into matter, which is similar to Einstein's Special Theory of Relativity. However, we also see that if we add-in energy, E, to an object, then, part of this energy, E_m, goes to form new matter and the rest remains as some form of energy, E_e, i.e. $E = E_m + E_e$. The value of E_m is given by $E_m = m_0 c^2 \{(1 + v / c)^2 - 1\}$, as Einstein himself showed with his thought experiment, that his famous mass/energy equation is independent of his set of relativistic transformation equations and thereby also applies to our relativistic theory. We also see that for $v \ll c$, i.e. non-relativistic situation, $E_m = 0$ and $E = E_e$, which is what we would expect.

Thus, we can say that when, $E \geq E_{min}$, then, we have $E = E_m + E_e$, with $E_m \neq 0$, and when $E < E_{min}$, we have $E = E_e$, with $E_m = 0$. Here, E_{min} is a new universal energy constant, which determines if E can turn into matter, or not. We will represent this new energy constant by the Greek symbol, K. Thus if $E \geq K$, then E can form matter, but if $E < K$, then E cannot form matter.

We can express our equation, $E = E_m + E_e$, as, $E \Leftrightarrow (E_m + E_e)$, which, logically, means, (a) $E \Rightarrow (E_m + E_e)$ and (b) $(E_m + E_e) \Rightarrow E$. We have already discussed the meaning of (a). From (b), we see that, when $E_m = 0$, then, $E_e \Rightarrow E$, which we already know from classical physics. But, if $E_m \neq 0$, then, (b) says that matter converts completely into energy, $(E - E_e)$, only. Thus energy can turn into matter and other forms of energy, but matter can only turn into energy. Thus, our equation $E = E_m + E_e$ represents a new general physical law, called, "The general law of conversion of matter and energy".

(II) Conclusions:

(1)   We see that unlike Einstein's $E = MC^2$, our equation, $E = E_m + E_e$, allows only part of E to convert into matter, with the rest remaining in some form(s) of energy.

(2)   If $E_e = 0$, then, $E = E_m$, which is the famous mass/energy equation by Einstein. Thus we see that the mass/energy equation by Einstein, which also describes the conversion of matter and energy, is a special case of the general law of the conversion of matter and energy.

(3)   From our expression, $E_m \Rightarrow (E - E_e)$, we see that, if, $(E - E_e) \geq K$, then the resulting energy, E_m, derived from the destruction of the matter, will immediately turn into new matter, E_m' and new form(s) of energy, E_e'.

(4)   From the equation, $E = E_m + E_e$, we see that, at the time of the origin of our universe, i.e. at $t = 0$, the virtual energy we discussed in my paper, "On the probability of origin of the universe and other matters", gives rise to not only real matter, M_R, but also real energy, E_e. This, E_e, I propose, will be in the form of the Unified Force Field. Thus, the energy that

would cause the matter to undergo the Big Bang, as discussed in that paper, will be created at the same time as the creation of that matter. As we have seen, in that same paper, the matter at $t = 0$ will be either ordinary matter or anti-matter but not both. However, the E_e, if $\geq K$, that is present can form both matter and anti-matter that can annihilate each other to give rise to just energy again. But, the resulting universe will be either ordinary matter or anti-matter dominant. In the case of our universe it is ordinary matter dominant.

References:

(1) Further conclusions based on the theory, "A relativistic theory based on the invariance of Newton's second law for motion and the constancy of the speed of light in vacuum". viXra: 1407.0105. viXra.org.

(2) On the probability of origin of the universe and other matters. Mustafa A. Khan. viXra: 1404.0431. viXra.org.

The equivalence of dark matter and dark energy and other thoughts.

If we represent all the dark matter in the universe by M, then:

1) $M \equiv \Lambda$, where Λ is the total dark energy in the universe. The Λ behaves differently on a mass, m, depending on the location of m within the universe.

2) Based on the theory, "On the consequences of a probabilistic space-time continuum", we have, $\Lambda = \sum_{m_v(T)} AGF(m_v)$ Here, AGF = the anti-gravitational field of visible matter m_v and $m_v(T)$ is the total visible matter in the universe.

3) Thus, $M \equiv \sum_{m_v(T)} AGF(m_v)$

4) If we have an object, m, near the edge of the universe, then there is will be an unbalanced force from Λ that will push the object farther away, i.e causing the expansion of the universe.

5) If we have an object, m, well within the universe, i.e. quite far away from the boundary, then the object will experience pressure from Λ that is equal on all the sides. However, the object does not collapse into a point because Λ also pervades within the object itself thereby causing an outward pressure on it and trying to tear it apart. These two forces, both of which are Λ, balance each other out and allow the object to maintain its form.

6) As an object moves closer to the edge of the universe, it will experience increasing unbalanced Λ with less force from Λ on side of the object facing the boundary. This will cause the object

to not only move with increasing speed or accelerate towards the boundary, but also causes it to start deforming. If our object is spherical, then the curvature of the object facing the boundary of the universe will be greater than the curvature on the other sides. The object will gradually start to flatten and will start to tear apart the closer it comes to the boundary of the universe. Initially, the other two forces, the electro-weak and the strong nuclear force will hold off against the force from Λ. Then the combined electro-weak and strong nuclear force will equal the force from Λ. After this point the force due to Λ dominates and starts to first deform it and then shred it into it's most elementary particles. The point where the force due to Λ is equal to the sum of the forces from electro-weak and strong nuclear force can become the starting point to unify the gravitational force and the other two forces. Hence, from this equality the equations for a unified field theory can be obtained.

7) The balloon analogy that is usually used to describe the shape of the universe with the 3D surface of the balloon representing the universe and the 2D spots on the surface representing the galaxies has to be wrong. This analogy is usually used to visually show the expansion of the universe and the apparent movement of the galaxies from each other in all the directions. This analogy shows that the more distant a galaxy from us the greater is the apparent speed with which it is moving away. But with the experiments, which proved the existence of dark energy, it is clear, that the more distant a galaxy the greater is the absolute speed with which it is moving away from us. It is not apparent but real. This means unlike the balloon analogy, which shows the universe to be finite and unbounded, the real universe actually has a boundary, which is accelerating away from us and from every other point within the 4D universe.

8) From # 6 we see that if we take $m_v(T)$ only the matter that is visible, then we will be grossly underestimating the total amount of $m_v(T)$, since, there is likely to be an immense amount of

m_v near the boundary of the universe that is an invisible visible matter.

9) From the theory, "A relativistic theory based on the invariance of Newton's second law for motion and the constancy of the speed of light in vacuum", we have $\Delta t' = \left(1 - \dfrac{\Gamma}{c}\right)\Delta t$. Hence, for light $\Gamma = c$ and $\Delta t' = 0$. This means eventhough relative to us light takes a finite non-zero amount of time to reach us, it takes zero amount of time relative to light itself to reach us. The theory also allows particles to move at any speed. Hence, if we have a particle that moves so fast that even though the speed is finite, but to us it "appears" to be "infinite", then we have a source carrying information that "appears" to us as being instantaneous and this solves the problem of quantum entanglement. This particle will transfer information from one end of the universe to the other in what "appear" to us to be instantaneously and allow two particles on the opposite sides of the universe to communicate with each other, in what "appears" to us to be instantaneous. However, since the universe is expanding at an accelerated rate, there will be a time when even this particle will appear to take a finite amount of time relative to us in transferring information between two particles on the opposite sides of the universe. At that time we will realize that the quantum entanglement is not instantaneous as we think it to be today but takes a finite amount of time or putting it in other words, we will then see that the quantum entanglement is not instantaneous. This situation is similar to the time when we once thought the speed of light to be infinite, until we measured it and found it to be finite. Then we realized that the speed of light being infinite was an illusion due to the distances we were considering.

References:

(1) On the consequences of a probabilistic space-time continuum. Mustafa A. Khan. viXra: 1401.0234. viXra.org.

(2) A relativistic theory based on the invariance of Newton's second law for motion and the constancy of the speed of light in vacuum. Mustafa A. Khan. viXra: 1406.0054. viXra.org.

An interesting, but not practically impossible, application based on the theories on gravity presented.

Here we will discuss an interesting theoretically possible and practically not impossible application based on the two proposed theories on gravitation. First, I like to point out that the theory; "On the consequences of a probabilistic space-time continuum" (PSTC) is independent of any particular theory of gravitation. It is applicable equally to Newton's Theory of Gravity (NTG) and to Einstein's General Theory of Relativity (GTR). In the paper on PSTC I discussed how it could be applied to NTG. Hence, the PSTC is also applicable to our theory, "On a general theory of gravity based on Quantum Interactions" (GQI).

Let us construct a vehicle, V, as follows:

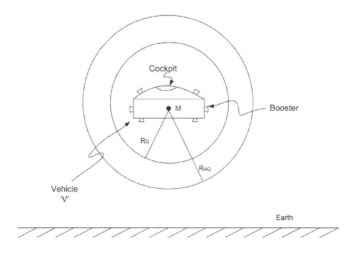

Here we have:

1) M = A mass, as per our GQI, producing a "gravitational field" (*GF*) around itself and enveloping V.

2) R_G = The maximum distance from M within which the *GF* is mostly attracting and which also envelops V.

3) R_{AG} = The minimum distance from M at which the *GF* is mostly repelling, i.e. "anti-gravitational field" (*AGF*), and which is quite larger than the size of V.

4) ($R_{AG} - R_G$) = The region in which the *GF* changes from attracting to repelling.

Now, regarding our V, we can draw the following conclusions:

1) The boosters function more like a rudder of a ship than as the primary power source for V's motion.

2) The *AGF* created by M is the primary power source that will move V.

3) For horizontal movement that is parallel to Earth, we have a situation that is similar to the Maglev that is used for the superfast trains. Of course, here, instead of a magnetic field we have *AGF*. Since there is no contact between the Earth and our V, our vehicle should also be able to achieve extremely high speeds due to the lack of friction. Also, our V will not be affected by the drag from Earth's atmosphere as the *AGF* will create an area that is almost a vacuum beyond R_{AG}. The pocket of air around V for $R \le R_G$ will be held by the *GF* of M and will be moving with V and thereby will not cause any drag on V either.

4) The vertical lift of V will also be quite fast and will get faster the farther away it is from the ground due to the decreasing gravitational field of the Earth. Our V may be able to achieve presently unimaginable speeds during vertical lift. Here also we need not be concerned about drag from the Earth's atmosphere.

5) The passengers in the cockpit will be exposed to tremendous accelerations that may not be suitable for them. This means we will need to have our V fitted with a mechanism where the *M* is slowly increased, thereby causing a gradual increase in the *AGF* of *M*. This will result in an acceleration of V that is not inimical to its passengers.

6) The *AGF* from *M* will provide a strong barrier against any attacks on V, as an attack by any material object will be strongly repelled by the *AGF*.

7) With appropriate engineering our V can do space travel at speeds that are presently unimaginable. By making the speed constant we can take advantage of the slowing of time allowed by either Einstein's Special Theory of Relativity or my own theory, "A relativistic theory based on the invariance of Newton's second law for motion and the constancy of the speed of light in vacuum". With this, destinations to distant stars within our Milky Way or even to other galaxies will not be beyond our grasp.

References:

(1) On the consequences of a probabilistic space-time continuum. Mustafa A. Khan. viXra: 1401.0234. viXra.org.

(2) On a general theory of gravity based on Quantum Interactions. Mustafa A. Khan. Mustafa A. Khan. viXra: 1405.0315. viXra. org.

(3) A relativistic theory based on the invariance of Newton's second law for motion and the constancy of the speed of light in vacuum. Mustafa A. Khan. viXra: 1406.0054. viXra.org.

The Sohraab-Hyder or SH set theory.

I) Introduction:

In my theory, "A relativistic theory based on the Invariance of Newton's second law for motion and the constancy of the speed of light in vacuum", we came across the equation $\Gamma = \dfrac{v'-v}{1-\dfrac{v}{c}}$. Later we found that the quantities v, v' and c were fictional as they were related to a fictional absolute inertial reference frame. However, the quantity, Γ, was found to be real as it was measurable and related to a real, non-absolute inertial reference frame. Thus the above equation tells us that an expression, $\dfrac{v'-v}{1-\dfrac{v}{c}}$, though containing fictional quantities can give a result that is real. Specifically, we see that the ratio of fictional quantities gives a result that is real. It is interesting to note that the laws of physics that we use in our everyday life and which form the basis of Newtonian Physics (NP) is actually based on fictional quantities or numbers. This means what we have considered to be real numbers throughout our human history are in fact fictional numbers. The non-fictional quantities or numbers we encounter in extreme cases such as in Quantum Mechanics (QM) and Relativity (R). The physical laws we encounter in QM and R actually use non-fictional numbers and describe phenomenon that is outside the realm of human senses. From Bohr's Correspondence Principle (CP) we find that the laws of QM and R are reducible to the laws of NP.

This means we are actually substituting non-fictional numbers with fictional numbers. Putting it in a different way, the human mind is better at perceiving fictional quantities/numbers than non-fictional quantities/numbers. Even though this notion is counter intuitive it is nevertheless true.

Using these ideas we can formulate a new type of set theory. We can say that all fictional numbers together with all the results of all the possible mathematical operations that can be done on them and all the equations that can be formulated using them (these, of course include all the laws of NP) with all the possible relationships between those equations form a set. This set we will call a Sohraab-Hyder or SH set. We can similarly form a SH set using the non-fictional numbers. Since, we have seen that a non-fictional number can be represented as the ratio of two fictional numbers, it is obvious that all the non-fictional numbers are contained in the SH set containing fictional numbers as we have also said that the fictional SH set, containing fictional numbers, also has as it's elements all the possible mathematical operations that can be done on them, which obviously includes the ratios of the fictional numbers. Now, since the non-fictional numbers are also elements of the SH set of fictional numbers, it means that all the elements of the SH set with non-fictional numbers are also present in the SH set containing fictional numbers. In other words, the SH set of non-fictional quantities is a subset of the SH set of fictional quantities. This means the SH set of fictional quantities not only contain the laws of NP, but also the laws of QM and R.

II) The Sohraab-Hyder or SH set:

We define a Sohraab-Hyder or SH set of ath order as one which consists of elements, designated by N^a, and which has the following properties: (1) the element $N^a = (N^b)'$, where N^b is the SH set of bth order from which we derive the SH set of ath order by putting the symbol for prime on the elements of the SH set of bth order. (2) N^a

represents not only any number that is derived from the numbers of N^b by putting the symbol for prime on them, but also represents all the possible results of all the possible mathematical operations that can be done on the elements of N^b with the symbol for prime placed on them. (3) N^a also represents all the possible results of all the possible mathematical operations that can be done on the elements of N^a. (4) N^a also represents all the possible equations that can be derived using all the elements of N^a.

At this point the reader will have the following questions:

1) We have defined a SH set using another SH set. This is tautological. I completely agree with this statement. However, to define a SH set in the broadest sense possible with the least number of words we can only use a definition that is tautological. This definition for a SH set suffices for now. As we explore the SH sets it will become clear to the reader, by the end, that a non-tautological definition can be formulated in a way that the reader can easily grasp the basic idea of a SH set. But if we start with the non-tautological definition now we will end up with a definition containing so many concepts that the definition will not only be too cumbersome for the reader, but the reader will not be able to grasp the basic idea of a SH set easily.

2) What do we mean by the order of a SH set? Remember that we have not defined what values a can take in N^a. We have not set a to be natural numbers, including zero, which will give us a linear order for the SH sets, such as N^0, N^1, N^2,.........and so forth. Our SH sets do not form a linear, stacking order, like a skyscraper, but form a tree with infinite branches with each branch giving rise to infinite branches and so on. The trunk of our SH set tree we represent by the Greek symbol Ω. This SH set, Ω, has as it's elements all the possible, whether real, fictional, non-fictional, non-(non-fictional) and so forth numbers together with all the possible results from all the possible mathematical operations between them and all the equations that can be formed

using them with all the possible results from all the possible mathematical operations that can be done between them and all the possible mathematical operations with all the other elements of the set. This SH set, Ω, thus contains within it the entire tree while at the same time forms the trunk of the above tree with infinite number of branches. I will discuss more about Ω later in this paper, but suffice it for now that Ω is the ultimate SH set and that there cannot be a superset of Ω, other than Ω itself, and like any set it is also it's own sub-set.

III) Properties of a SH set:

1) Just for the sake to be complete, we can easily see that there are an infinite number of SH sets, with both finite, which includes zero number of elements, and infinite number of elements.

2) It is not necessary that $N^b \subset N^a$ where $N^a = \left(N^b \right)'$. The two SH sets we discussed in the introduction are an exception due to the fact that we know the relationship between the real and fictional numbers, namely, a real number is a ratio of two fictional numbers. Other than these two SH sets we do not have any knowledge about any relationship between N^a and N^b in general, except for $N^a = \left(N^b \right)'$.

3) Any SH subset, including Ω, is a subset of itself, i.e. $N^a \subset N^a$.

4) A SH set can have zero to infinite number of elements.

5) The null SH set, $\{\}$, is subset to every other SH set including itself.

6) The next order SH set of a null SH set is a null SH set.

7) The Ω SH set is both it's own subset and it's own superset, i.e. $\Omega \subset \Omega$ and $\Omega \supset \Omega$. This we represent by $\Omega \equiv \Omega$.

IV) Further thoughts on Ω :

As promised I like to discuss Ω further. As was said before, Ω is the ultimate SH set. It is quite obvious that the number of elements in Ω is infinite. We will represent an element of Ω by the symbol N^∞ , i.e. Ω is a SH set of infinite order. Of course, there are an infinite number of SH sets with an infinite number of elements, but we will restrict the use of the symbol N^∞ to represent an element of Ω only. It is easily seen that if we try to create a SH set from Ω, i.e. $\left(N^\infty \right)'$ we end up with N^∞ only. This is because the symbol ∞ does not represent a quantity or a number, but a concept and therefore the element $\left(N^x \right)'$ is also already an element of Ω. Suppose we form a Ω' from Ω. Then the elements of Ω', which we will represent by $\left(N^\infty \right)'$ are not contained in Ω. But this contradicts the definition of Ω as was given before when we discussed the meaning of the order of a SH set. Hence, $\left(N^x \right)'$ must also be an element of Ω. This means $\Omega' \subset \Omega$. By the same logic if we have $\left(\Omega' \right)'$, then it will also be a subset of Ω. Thus Ω has no superset other than itself. To use our tree analogy, we can say that Ω is the trunk of all the infinitely possible trees or to put it in another terms, the tree whose trunk is Ω is the only tree that can possibly exist. In the section, "Discussion and Conclusions", I will show that Ω can not only possibly exist, but that it must exist. Also, that Ω has always existed and will always exist. It is self-sufficient. It created itself in pre-eternity and it cannot be destroyed, even by itself, i.e. Ω can never be reduced to a null SH set.

V) Discussion and Conclusions:

At this point the reader may well ask as to why do we need yet another set theory? The answer to this question, as the reader will find out, lies in the following discussion and conclusions:

1) It is clear that all the possible sets of the current set theory that has mathematical quantities, as their elements are also SH sets.

2) As we have seen in the introduction, time, either Newtonian (NT) or Einstein (ET) are also elements of a SH set. This can be generalized to say that any kind of time, even non-NT and non-ET are also an element of a SH set.

3) There must exist an infinite number of SH sets that has as their elements mathematical quantities that form logically consistent algebras.

4) Since Descartes showed that geometry can be expressed using algebraic equations, we can see that there must exist an infinite number of SH sets with elements that are algebraic equations and which form a logically consistent geometries.

5) From the set theory we know that multiple copies of an element(s) do not change the set. For example, the set $\{0,1\}$ is equivalent to the set $\{0, 1,1,1...$infinite times$\}$. Hence the set Ω is the same even if it has an infinite number of each of its infinite elements. This means it has infinite number of infinite kinds of "time" with infinite number of equations containing these infinite kinds of "times". Other than the equations that has "time" which does not move, all the other infinite equations are continuously producing and destroying the infinitely various number of elements of Ω. However, even the infinite number of equations, that destroy the other elements, that contain "time" that moves at an infinite rate will never be able to destroy all the elements of Ω to reduce it into a null SH set because they will take an infinite amount of time to destroy all the infinite number of elements present in Ω. Besides, Ω also has equations that are creating elements at an infinite rate using the infinite kinds of "times". Hence, eventhough Ω is infinitely dynamic it is the same at any given moment of the infinite kinds of "times". One can take the analogy of our Sun. Though it appears to be unchanging at any moment of time, we know that it is extremely dynamic and undergoing violent and rapid changes at every moment in time. This proves that Ω has always existed and will always exist and never changes. Since $\Omega \equiv \Omega$, it is self-sufficient.

6) The existence of Ω automatically guarantees the existence of all the infinite number of SH sets that are subsets of Ω. This means the Ω guarantees the existence of the SH set that contains the physical laws (or equations) that operate within our universe. Also, Ω automatically guarantees the existence of infinite number of SH sets that will give rise to infinite number of universes with infinitely different and logically consistent set of physical laws including infinite number of universes with the same physical laws as those in our own universe.

7) From # 6 above we can easily see that Ω automatically guarantees the existence of an infinite number of SH sets that have finite number of elements and which constantly change due to one of the infinite kinds of "time" being present in them and the equation(s) which are also present in them. These special type of SH sets we will call "mathematical cells".

8) These mathematical cells can have different properties depending upon their elements. Unless a mathematical cell has the kind of "time" that does not move, they are necessarily dynamic units. They can, (1) create and/or destroy elements within them, (2) take in elements from around them if they have the appropriate equation(s) to allow such a process, (3) they can form bonds between each other or with other mathematical cells to give rise to organs and organisms from the simplest to the most complex. These are all, of course, mathematical organs and organisms.

9) The Ω automatically guarantees the existence of an infinite number of mathematical cells that has the property that makes them evolve and thereby makes the organism built with them to also evolve. In short Ω automatically guarantees evolution of mathematical organisms.

10) Since we know that the organic cells use the same physical laws as the rest of the non-organic universe, we can conclude that some of the infinite numbers of our mathematical cells are equivalent to the organic cells. This means the entire organic universe is nothing more or less than a mathematical universe. This also means Ω automatically guarantees the creation and

evolution of the entire organic universe, which also includes us human beings. Also, we see that Ω guarantees the evolution of organisms as propounded by Charles Darwin in his famous book, "On the origin of species".

11) Finally, as we have learned from the computers that any and every mathematical quantity (which includes equations) is reducible to a string of 0's and 1's, we can conclude that this entire universe, which includes both living and non-living entities, are ultimately strings of 0's and 1's. Hence, the creators of the movie "Matrix" were right when they showed that everything is a matrix of 0's and 1's. Here, we have mathematically proved that their concept/ intuition was/is correct!

12) In another paper I will delve deeper into the concept of the mathematical cell and construct, "The Mathematical Cell Theory".

References:

(1) A relativistic theory based on the Invariance of Newton's second law for motion and the constancy of the speed of light in vacuum. Mustafa A. Khan. viXra: 1406.0054. viXra.org.